OCEAN MAMMALS

ORCAS

by Julie Murray

Cody Koala
An Imprint of Pop!
popbooksonline.com

Hello! My name is Cody Koala

This book is filled with videos, puzzles, games, and more! Scan the QR codes* while you read, or visit the website below to make this book pop.

popbooksonline.com/orca

*Scanning QR codes requires a web-enabled smart device with a QR code reader app and a camera.

abdobooks.com

Published by Pop!, a division of ABDO, PO Box 398166, Minneapolis, Minnesota 55439.

Printed in the United States of America, North Mankato, Minnesota.

042025
082025

THIS BOOK CONTAINS RECYCLED MATERIALS

Cover Photo: Getty Images
Interior Photos: Shutterstock Images; Getty Images
Editors: Elizabeth Andrews and Grace Hansen
Series Designer: Candice Keimig

Library of Congress Control Number: 2024948405

Publisher's Cataloging-in-Publication Data

Names: Murray, Julie, author.
Title: Orcas / by Julie Murray
Description: Minneapolis, Minnesota : Pop!, 2026 | Series: Ocean mammals | Includes online resources and index
Identifiers: ISBN 9781098247751 (lib. bdg.) | ISBN 9781098248291 (ebook)
Subjects: LCSH: Orca--Juvenile literature. | Killer whale--Juvenile literature. | Whales--Behavior--Juvenile literature. | Marine mammals--Juvenile literature. | Aquatic mammals--Juvenile literature. | Sea animals--Juvenile literature.
Classification: DDC 599.536--dc23

Table of Contents

Chapter 1

Pod Life

Orcas are **mammals**. They are also known as killer whales. But they are not whales. They are the largest members of the dolphin family! They are social animals and live in groups called pods.

An orca pod can have 5 to 50 members.

Watch a video here!

Orcas are found swimming in oceans all around the world. They mainly swim near the shore. Some orcas **migrate** long distances. They can swim up to 28 miles per hour (45kph).

Where Orcas Live

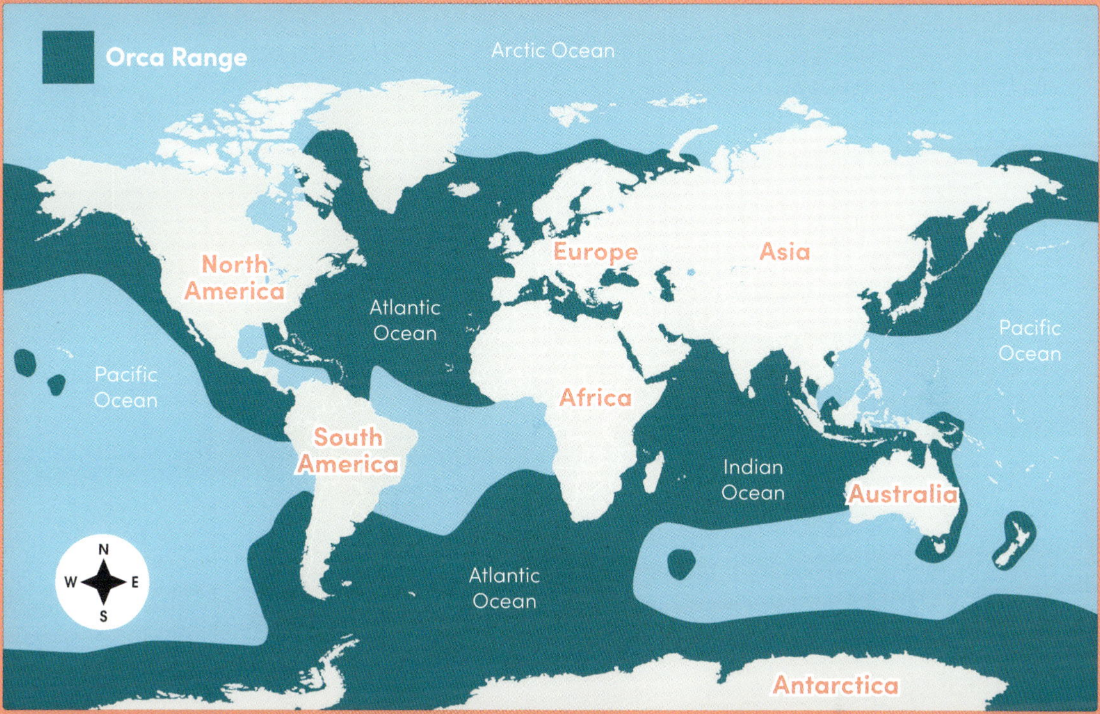

Orcas have large bodies. They are 25 feet (7.6m) long and can weigh over 10,000 pounds (4,500kg)!

Their smooth skin is black and white. The colors help them hide in the ocean.

pectoral
fin
dorsal
fin

Orcas have large, rounded pectoral fins. These fins can be 7 feet (2.1m) long. An orca's dorsal fin is black and can stand 6 feet (1.8m) tall. Its tail fins, called flukes, are white underneath.

Orcas have sharp, curved teeth for hunting. They breathe through a blowhole on the top of their head.

An orca sleeps with one eye open. Half its brain shuts down to sleep. The other half stays awake to control its breathing.

The blowhole stays closed underwater. It opens when the orcas reach the water's surface.

Chapter 2

Communication

Orcas use special sounds to **communicate** with one another. Each pod has its own sounds. These sounds help orcas **recognize** other pod members.

Orcas use **echolocation** to communicate, travel, and find food.

Learn more here!

Orcas also slap their fins against the water to communicate. They might also come to the surface just to have fun!

Spyhopping is when an orca comes slowly out of the water to check out its surroundings.

Chapter 3

Food

Orcas are one of the ocean's best **predators**. They eat up to 500 pounds (227kg) of food a day! They often hunt in groups. They eat fish, seals, and sea lions. They also eat whales and sharks.

Explore links here!

Chapter 4

One Calf

Female orcas give birth every three to five years. They have one baby, called a calf, at a time. The calf is 8 feet (2.4m) long and weighs 450 pounds (204kg) at birth.

Male orcas can live for 60 years. Females can live for 90 years!

Complete an activity here!

Making Connections

Text-to-Self

Would you want to see an orca in real life? Why or why not?

Text-to-Text

Have you read books about other ocean mammals? How are they similar to or different from orcas?

Text-to-World

Orcas live in groups. Can you name any other animals that live in groups? What is helpful about living in a group?

Glossary

communicate – to share information with others.

echolocation – using sound to sense where things are.

mammal – a warm-blooded animal that produces milk for its young and is usually covered with hair.

migrate – to travel to a new area, often seasonally.

predator – an animal that hunts other animals for food.

recognize – to identify someone or something.

Index

Online Resources

popbooksonline.com

Thanks for reading this Cody Koala book!

This book is filled with videos, puzzles, games, and more! Scan the QR codes* while you read, or visit the website below to make this book pop.

popbooksonline.com/orca

*Scanning QR codes requires a web-enabled smart device with a QR code reader app and a camera.